GCSE 9-1

geography

AQA

Fieldwork

Series editor
Bob Digby   **David Holmes**

OXFORD
UNIVERSITY PRESS

Contents

Fieldwork and AQA GCSE Geography

This book is written to help prepare you for the fieldwork section in Paper 3 of AQA's GCSE Geography specification. It's a workbook with space to write, and practice exam-style questions

Preparation

You should have undertaken the equivalent of two days' fieldwork during your GCSE course, one in physical geography, and one in human.

> Each gives you the chance to collect primary data (data you collect yourself) on topics linked to content you've learned in Units 3.1 and 3.2.

> Questions in the exam won't be specific about where you went or the topic you studied, because every school can choose what it likes. Expect open-ended questions which refer to *'a place in which you carried out fieldwork'*, or *'For your physical geography enquiry, explain how …'*.

How fieldwork is assessed

Fieldwork is assessed in Section B of Paper 3. The exam lasts 1 hour 15 minutes, split between the Issue Evaluation in Section A (worth 37 marks, including 3 for SPaG), and Fieldwork in Section B (worth 39 marks, including 3 for SPaG).

> You'll have 38 minutes to do Section B! It's pressurised, but you'll cope if you're well prepared. That's where this book can help.

In the exam, you'll find two types of fieldwork questions:

> **familiar** fieldwork (that you've done). These questions ask about skills you used, and will expect you to think critically about what you did (for example, improving your fieldwork).

> **unfamiliar** fieldwork (that others have done). These questions ask you to apply what you've learned to new situations (for example, applying skills).

Using this book

Fieldwork for AQA's GCSE Geography specification has six strands. These are:

> 1 Devising questions for geographical enquiry
> 2 Selecting, measuring and recording data
> 3 Selecting appropriate ways of processing and presenting fieldwork data
> 4 Describing, analysing and explaining fieldwork data
> 5 Reaching conclusions
> 6 Evaluation of geographical enquiry

These strands form the six chapters in this book. The exam will contain questions about any, or all, of these strands.

Using each chapter

Each chapter in this book consists of ten pages.

> • The first four pages **prepare** you for the exam. Each contains activities designed to help you understand that strand, and prepare you for exam questions.
> • The next six pages consist of exam-style questions to use for **practice**. The mark schemes are online at www.oxfordsecondary.co.uk/aqa_gcse_geog.

Guiding you through each chapter

Every chapter includes:

> • A 'Big Idea' which sums up the chapter.
> • 'Over to you' activities to help you understand the chapter.
> • 'Exam hints' which give you specific guidance about tackling exam questions.
> • 'Get Exam Ready' which helps you to apply the fieldwork that you did to each chapter.

Guided answers are available on the Oxford Secondary Geography website:
www.oxfordsecondary.co.uk/aqa_gcse_geog

Please note this revision guide has not been written or approved by AQA. The answers and commentaries provided represent one interpretation only and other solutions may be appropriate.

Geographical Enquiry Strand 1

1 Thinking of a question, title or hypothesis for geographical enquiry.
2 Using a concept, process or theory in your enquiry.
3 Using primary and secondary data for your enquiry.
4 Identifying and reducing potential fieldwork risks.

 Getting started

Why fieldwork?

Geography and fieldwork go hand in hand. Geographers study the world's physical surroundings and how people use space, and the issues that can arise from this. The core of Geography is the real-world study of real places and people. To become a geographer and to think like one, you need to do real-world fieldwork! That's where this book will help you.

But what sort of places do geographers visit to carry out their fieldwork? And what do they do when they get there? Your teacher will guide you in how to carry out most fieldwork tasks. But this first section will help you think about how geographers start, by developing an enquiry question, title, or focus.

 Big idea

Starting your enquiry can be difficult. Reading your textbook and researching online can help you to think of ideas. Then you can start to plan!

> We went to Start Bay in Devon to explore problems caused by longshore drift. This was important as longshore drift moves material along the beach. In 1984 John Jackson came to Hall Sands and took 650 000 tonnes of sand off the beach which means some houses are not protected. Now the beach is thinner on one side compared to the other.

Figure 1 An extract from a student's introduction and enquiry focus

Read the passage in Figure 1.

a) What do you think is the focus for the student's enquiry?

_____

_____

b) How clear is it?

_____

_____

c) How could the student have improved their focus?

_____

_____

1 Thinking of a question, title or hypothesis for geographical enquiry

It's easy to be over-ambitious! Choose something which is small scale, and workable, as it is more likely to give you reliable and accurate data. It can be a question, or a hypothesis (a prediction that you want to test) or just a plain title.

Choosing a good question, title or hypothesis can depend on:

- the place – is it a good example of what you want to study? Is it accessible?
- scale – don't choose an area or environment that is too large to study.
- data – can you measure what you need to?

 Exam hint!

It doesn't matter whether you use a question, title or hypothesis. Just make sure you can write down the aim of your fieldwork!

 **Over to you**

Figure 2 shows some examples of fieldwork enquiries. Complete the table to show whether you think each is a good focus for a fieldwork enquiry or not. Give your reasons.

Example of titles	Suitable focus?	Reason
Is hard engineering working in Cromer?	Yes / No	
A study of regeneration in east London	Yes / No	
Will the River Onny will get bigger downstream?	Yes / No	
What are the advantages and disadvantages of tourism on the seafront at Blackpool?	Yes / No	

Figure 2 A good focus for fieldwork, or not?

2 Using a concept, process, or theory in your enquiry

Your enquiry could be based on something you've already studied such as:

- a concept (e.g. sustainability – see Figure 3)
- a process (e.g. how the size of pebbles on a beach changes with longshore drift)
- a theory (e.g. how a river channel changes as it moves around a meander).

Figure 3 Investigating a concept. How sustainable is this as a place in which to live?

Over to you

For each title in Figure 2 on page 5, name one concept or theory on which each title is based. The first one has been done for you.

a) **Title:** Is hard engineering working in Cromer? **Concept / theory:** coastal management and the debate between hard and soft engineering.

b) **Title:** A study of regeneration in east London. **Concept / theory:** _____

_____

c) **Title:** Will the River Onny will get bigger downstream? **Concept / theory:** _____

_____

d) **Title:** What are the advantages and disadvantages of tourism on the seafront at Blackpool?

Concept / theory: _____

3 Using primary and secondary data for your enquiry

Primary data are data which you collect yourself, or as part of a group.
Secondary data are collected by someone else. Secondary data can be important in helping you to research an idea or focus.

Over to you

Complete the table in Figure 4 with sources of primary and secondary data that you could collect for the enquiry titles listed.

Title	Two sources of primary data	One source of secondary data
Has urban regeneration improved the environment in Leeds city centre?	1 2	
Has river management reduced flooding for people who live near the River Nene in Peterborough?	1 2	

Figure 4 Thinking about sources of primary and secondary data

4 Identifying and reducing potential fieldwork risks

Questions relating to possible risks when carrying out fieldwork, and how it can be reduced, could form part of your exam, so you need to think about risk assessment. This means:

1 identifying hazards or risks
2 deciding who might be harmed, and how
3 deciding how to reduce the risks.

Over to you

Complete the risk assessment in Figure 5 for *either* your physical *or* human fieldwork. Identify two risks or hazards. Levels of risk go from 1 (low) to 5 (very high). An example has been done for you.

Risk / hazard	Level of risk	Who might be harmed?	How to reduce the risk
Sand dunes – injury through climbing and jumping	*2*	*Students*	*Students are warned of hazard and told not to jump / play in dunes*

Figure 5 A risk assessment

Get exam ready

In the exam you'll need to name the focus of your enquiry.
Practice by completing the details here.

1a) The topic I studied for my **physical** fieldwork was: _____

b) The place I went to was: _____

c) The data I collected included: _____

_____

2a) The topic I studied for my **human** fieldwork was: _____

b) The place I went to was: _____

c) The data I collected included: _____

_____

Study Figure 1a, a photograph of a stretch of coastline, and Figure 1b, a photograph of a rural upland area.

Figure 1a A stretch of UK coastline, with a beach

Figure 1b Part of an upland rural area

1.1   Identify **one** question or hypothesis that could be investigated using fieldwork in **each** of the areas shown in **Figures 1a** and **1b**.

[2 marks]

Question or hypothesis for Figure 1a:

_____

_____

Question or hypothesis for Figure 1b:

_____

_____

Study Figure 2, a cliff with different rock types in north-east England.

Figure 2 A cliff with different rock types in north-east England

1.2   Explain how a secondary source (such as the photo in **Figure 2**) can help you to develop a suitable question or hypothesis for a geographical enquiry.

[2 marks]

_____

_____

_____

_____

Figure 3 is a diagram showing how land-use and the age of buildings varies in a city.

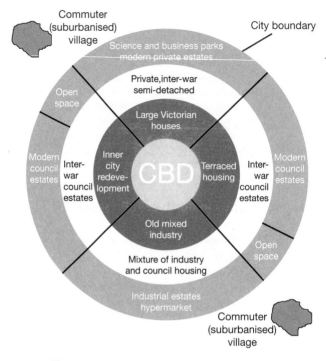

Figure 3 Land-use and building age in a city

1.3 Explain **one** way that a diagram like **Figure 3** could be used to help plan a fieldwork investigation.

[2 marks]

_____

_____

_____

_____

1.4 Explain **two** problems of using a diagram such as **Figure 3** in order to plan a fieldwork investigation.

[4 marks]

Problem 1: _____

_____

_____

_____

Problem 2: _____

_____

_____

Study Figure 4, a photograph of students carrying out fieldwork in a woodland environment.

Figure 4 Students carrying out fieldwork in a woodland environment

1.5   Explain **two** possible risks in the area shown in **Figure 4**.

[4 marks]

Risk 1: _____

_____

_____

_____

Risk 2: _____

_____

_____

_____

1.6   Explain how **one** risk you identified from **Figure 4** could be managed.

[2 marks]

_____

_____

_____

_____

Geographical Enquiry Strand 1

Exam-style questions: Familiar Fieldwork

2.1 State the title of your fieldwork enquiry in which **human** geography data were collected.

Title of fieldwork enquiry:

_____

Explain why it was a suitable title for a fieldwork enquiry.

[2 marks]

_____

_____

_____

_____

2.2 Explain **one** factor you considered when selecting a suitable question / hypothesis for your **human** geography enquiry about the place.

[2 marks]

_____

_____

_____

_____

2.3 State the title of your fieldwork enquiry in which **physical** geography data were collected.

Title of fieldwork enquiry:

_____

Explain why you selected any **two** secondary sources you used as part of your fieldwork enquiry.

[6 marks]

_____

_____

_____

_____

_____

_____

_____

_____

_____

_____

_____

_____

2.4   With reference to the planning and design in **one** of your enquiries, assess how helpful a geographical
theory or concept(s) were in developing your enquiry.

Title of fieldwork enquiry:

_____

[9 marks]
[+ 3 SPaG marks]

_____

_____

_____

_____

_____

_____

_____

_____

_____

_____

_____

_____

_____

_____

_____

_____

Geographical Enquiry Strand 2

1 Knowing the differences between primary and secondary data.
2 Choosing suitable physical and human data.
3 Using different sampling methods.
4 Justifying your methods of data collection.

1 The differences between primary and secondary data

Primary and secondary data lie at the core of fieldwork.

- **Primary data** are data collected by you or in a group, first-hand.
- **Secondary data** are data collected by someone else, and can include data collected by other groups, and published data. There are lots of different types of secondary data – including online reports, books, newspapers, photos or tables of data.

 **Big idea**

It's important to get the right data and sampling method. Good data should result in a more reliable conclusion.

Over to you

Choose two types of secondary data which will be relevant to your enquiry.

Secondary data 1 _____

Secondary data 2 _____

Complete the table to help you think about how good the data might be.

	Secondary data 1	Secondary data 2
Who produced it?		
Why was it published?		
How old are the data?		
Do other examples of secondary data give a different view or opinion?		
How far can the source be trusted?		

2 Choosing suitable physical and human data

The key to successful fieldwork is collecting data that can help you in your enquiry.

✎ **Over to you**

Figure 1 is a list of fieldwork focuses, both physical and human. Select examples of the type of data that could be collected, using the ideas in the box below the table (some are distractors only!) to complete it. You can use each more than once if necessary. Think carefully about some of the processes that might be going on.

Fieldwork focus	Appropriate fieldwork data
Success of a new shopping centre	*Pedestrian counts*
Changes in a river channel downstream	
Shape of a storm beach	
People's attitudes to a new coastal flood scheme	
Effect of tourists on honeypot sites	

Figure 1 Which data suit which focus?

river velocity (or speed)	pebble size and shape		infiltration rates
footpath erosion		beach gradient (profile)	traffic counts
interviews	environmental quality surveys		wind speed

3 Using different sampling methods

A **sample** is a set of data which you collect. **Sample size** is important because it helps to know whether the data you collected were representative of the place you studied. The larger your sample, the more reliable your conclusions will be. You can use one of four sampling methods.

- **Random** – samples are chosen randomly, e.g. every item, pebble, or person etc has an equal chance of being selected.
- **Systematic** – means having a system to collect data, e.g. every 10 cm across a river, or every 10th person surveyed on a street.
- **Stratified** – means a sample made up of different parts, e.g. selecting different pebble sizes from a river so that you include the whole range of sizes found there. The same goes for gender or age groups in a town.
- **Opportunistic** – which is where you more or less have to take the sample as it is, for example if you are carrying out a questionnaire in a quiet village where few people are about.

 Exam hint

Make a mnemonic to help you remember the four sampling methods from their initial letter, e.g.:
Random
Opportunistic
Systematic
Stratified – to make
ROSS (or create your own!)

Over to you

Complete the table below with the sampling method you would choose, and reason, to collect the data needed.

Data collection proposal	Type of sampling method	Reason for choosing this
Measuring sediment along a beach profile.		
Carrying out a questionnaire in a busy tourist hotspot.		
Carrying out Environmental Quality Surveys in a small town which has four very varied housing estates.		

4 Justifying your methods of data collection

It is often hard to know why a particular method of fieldwork data collection was chosen. A student, for example, might find they have to justify the following:

- Why they sampled every 30 m along a main road out of town.
- Why they asked 50 people as part of a questionnaire survey.
- Why they measured the depth of the stream at five places across the channel.

Over to you

Figure 2 shows two extracts from student answers to the exam question:

Justify one primary data collection method used in your human geography enquiry. (3 marks)

Student A

In Taunton we carried out a town health check by counting the number of vacant properties along the high street. This was used as the number of shop vacancies can be linked to the economic strength of a place. It was easy to count vacant shops, which should improve our reliability.

Student B

We used several methods to collect data about shopping quality in Leeds city centre. This included a questionnaire (this was quick and easy), plus an environmental quality survey and a traffic survey. Our results showed that it was a good place for shopping and there were no problems with our methods.

Figure 2 Two student exam question answers

Mark the answers in Figure 2 using the mark scheme in the table. Which answer is better?

	Student A answer	Student B answer
One method with detailed justification = 3 marks		
Some justification linked to the aims of the investigation = 2 marks		
Statement(s) with little justification = 1 mark		

✓ **Get exam ready**

In the exam, be ready to name the data collection methods you used in your enquiry.
Do this by completing the details below.

1 One example of primary *quantitative* or *qualitative* data I collected was: _____

_____

I collected this to help me find out _____

_____

2 One example of secondary data I collected was: _____

_____

I collected this to help me find out _____

_____

3 A sampling method I used was: _____

which I used to collect data on _____

Geographical Enquiry Strand 2

Exam-style questions: Unfamiliar Fieldwork

Study Figure 1, which includes examples of primary and secondary data.

Figure 1 Examples of primary and secondary data. Item 3 shows information about the Index of Multiple Deprivation for south Devon, colour coded according to the ranked score. Red is worst, green is best.

1.1 Identify the **two** pictures that show the collection of primary data by putting a cross in the correct boxes.

[2 marks]

Picture 1	Picture 2	Picture 3	Picture 4

Study Figure 2, a sketch map of five sites used to collect primary fieldwork data about a river.

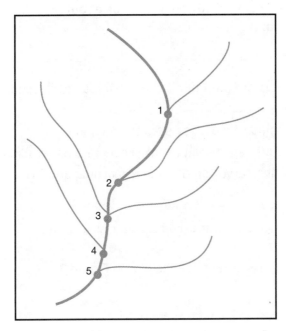

Figure 2 Sites used to collect primary fieldwork river data

1.2 Explain **two** reasons why students might have chosen a stratified sampling approach, as shown in **Figure 2**.

[4 marks]

Reason 1 _____

_____

_____

_____

Reason 2 _____

_____

_____

_____

1.3 A group of students wants to find out about the health of a town centre.

Explain **one** reason why they planned to collect footfall (pedestrian counts) as part of the investigation.

[2 marks]

_____

_____

_____

_____

Geographical Enquiry Strand 2

Exam-style questions: Unfamiliar Fieldwork

Study Figure 3, an extract from a student's fieldwork notebook about how they plan to collect data in an urban area.

- I will count people at 50m intervals along three roads coming out from the town centre.
- I will count people at the survey sites for 3 minutes.
- To make the count easier I will count people who are walking towards me.
- I will carry out the survey on a Monday morning and Friday lunchtime.

Figure 3 A plan to collect data in an urban area

1.4   Explain **two** potential problems associated with the plan in **Figure 3**.

[4 marks]

Problem 1 _____

_____

_____

_____

Problem 2 _____

_____

_____

_____

1.5   Explain **one** way in which the plan in **Figure 3** could be improved.

[2 marks]

_____

_____

_____

_____

Study **Figure 4**, part of a student's questionnaire to investigate people's shopping habits.

Shopping habits questionnaire

SEX　　　male ☐　　female ☐　　　　　　　　AGE　☐　☐　☐　☐

　　　　　　　　　　　　　　　　　　　　　　18-25　26-35　36-59　60+

WHERE DO YOU LIVE?　　☐　　　　　　☐　　　　　　☐

　　　　　　　Less than a mile from　1-4 miles from　over 4 miles from
　　　　　　　the shopping centre　the shopping centre　the shopping centre

HOW OFTEN DO YOU　　☐　　　☐　　　☐　　　☐
USUALLY GO SHOPPING?　Daily　2-3 times a　Once a week　Less often
　　　　　　　　　　　　　week

WHEN DO YOU USUALLY　☐　　　☐　　　☐　　　☐
DO YOUR SHOPPING?　Weekday　Weekday　Weekday　Weekend
　　　　　　　morning　afternoon　evening

WHAT MEANS OF TRANSPORT
DO YOU USE MOST OFTEN　☐　　　☐　　　☐　　　☐
WHEN YOU GO SHOPPING?　Car　Bus　Walk　Other

Figure 4 Extract from a questionnaire on shopping habits

1.6　Identify **two** possible problems with the design of the questionnaire in **Figure 4**.

[2 marks]

Problem 1 _____

_____

Problem 2 _____

_____

1.7　Explain **one** aspect of design that this student should consider when using the questionnaire with people in the street.

[2 marks]

_____

_____

_____

_____

Geographical Enquiry Strand 2

Exam-style questions: Familiar Fieldwork

2.1   State the title of your fieldwork enquiry in which **human** geography data were collected.

Title of fieldwork enquiry:

_____

Justify **one** method of data collection that you decided to use.

[2 marks]

Method used:

_____

Justification:

_____

_____

_____

_____

2.2   State the title of your fieldwork enquiry in which **physical** geography data were collected.

Title of fieldwork enquiry:

_____

Justify **one** method of data collection that you decided to use.

[2 marks]

Method used:

_____

Justification:

_____

_____

_____

_____

2.3   Justify the sampling techniques used in your **physical** geography enquiry.

[6 marks]

_____

_____

_____

_____
_____
_____
_____
_____
_____
_____
_____
_____

2.4   Referring to **one** of your enquiries, assess the extent to which you were successful in collecting primary data.

Title of fieldwork enquiry:

_____

[9 marks]
[+ 3 SPaG marks]

_____
_____
_____
_____
_____
_____
_____
_____
_____
_____
_____
_____

Continue your answer on another piece of paper.

Geographical Enquiry Strand 3

1 Knowing that there are a range of methods for presenting your data.
2 Selecting the most appropriate methods of presentation.
3 Describing, explaining and adapting your presentation methods.

1 Knowing a range of methods for presenting your data

As part of your enquiry you will collect primary fieldwork data which you'll share with others. This allows you to increase the size of your data sample. If the data come from different places, they can also help you to make comparisons.

Next you'll need to decide how to present your data. There are five main methods available, shown in Figure 1.

Big idea

Choose the best geographical way of presenting your data that you've collected. For example, don't just create graphs – show them on a map!

Method	When you would use it	Advantages
Maps / cartography		Makes it easier to compare patterns at different locations.
Table(s) of data		Can help to identify anomalies (any data which look unusual).
Graphs		There is a wide range of graphs and charts available.
Photos and field sketches of features		Helps you to pick out what's important using annotations.
GIS		Can be used to map several sets of data (e.g. questionnaire results and pedestrian flows), complex data (e.g. census data), or aerial photos.

Figure 1 When to use different methods of presenting data

The cells in the column headed 'When you would use it' in Figure 1 have been left blank.
Decide which of the following statements should go in each cell, and complete the table.

a) To present raw data that you and your group collected.
b) To show change over time e.g. coastal erosion, or changes to a town.
c) To show data and patterns clearly – easier to read than a table of data.
d) Used to show locations and patterns.
e) When you want to look at features in detail.

2 Selecting the most appropriate methods of presentation

 Exam hint

Always make sure your graphs or maps are accurately presented, and have a title, scale, labelled axes (for a graph) and key (for a map)!

The word 'appropriate' refers to the best way to present information that has been collected as part of an enquiry. There is no point in using a graph or cartographical technique if it is not helpful in understanding the data. The person looking at the method of data presentation needs to understand what it is showing.

Over to you

1   Look at the type of information in the table in Figure 2. Link each example to the best method of presenting it. An example has been done for you.

Type of information	Example of fieldwork data	The best method of presenting this information
Data showing changes over time	Changes in pedestrian flows over a 6 hour period	A land use map
Data collected to show spatial variation	Map of cafes in a town centre	A histogram
Data collected to show spatial variations as movement or flows	Frequency of traffic movements around a city	A line graph
Data that has a compass direction or orientation	Direction of wind over a year (as frequencies with an angle)	Proportional flow arrows
Numerical data for locations that has different categories on the x-axis	Visitor numbers (frequency) for people living in place X	A bar chart
Numerical data for locations that is continuous	Attitudes towards tourism of people in different age categories	A circular graph and rose diagram

Figure 2 When to use different methods of presenting data

Over to you

2 Complete the table below to show the best methods of presenting the data collected.

Enquiry title	Methods of collecting data	Best method of presenting data
How do pebbles change in size along a beach?	1 Pebble sizes at five locations.	1
	2 Slope of a beach at one location.	2
Do rivers get faster as they flow downstream?	1 Recording time taken for a float to travel 5 metres at ten places.	1
	2 Gradient (slope) of the river at the same ten places.	2
Do people prefer to shop in city centres or out of town?	1 Questionnaire asking people's preferences about where to shop.	1
	2 How people travel to shops	2
Do tourists bring benefits or problems to rural areas?	1 House prices in a village.	1
	2 Where second homes are located in a village.	2

③ Describing, explaining and adapting your presentation methods

In the exam you could be asked about data presentation. Questions could ask you to:

- describe a presentation method that you've used
- explain why the method you chose is appropriate, or possibly inappropriate
- suggest changes (adaptations) to the methods you chose to improve their accuracy and appropriateness.

Over to you

Read the two student answers to the question below.

Study Figure 3. Suggest two or more ways in which the data presentation technique could be adapted so it is a) more accurate, and b) more appropriate.

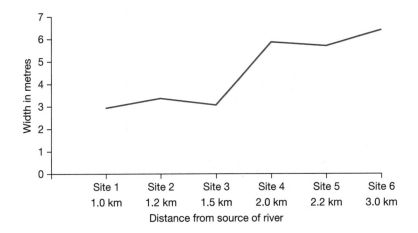

Figure 3 Graph to show the changing width of the River Fordern downstream from its source

Student A

> I would change it to a histogram as the sites will be separated so the data are not continuous.

Student B

> I would change it to a histogram as the sites are continuous

Which is the stronger answer? Why? _____

_____

_____

Get exam ready

Remember! In the exam you won't be asked to simply describe your data presentation methods. Expect questions which ask you to explain or justify why you used certain methods.

1a) One method I used to present my primary *quantitative* or *qualitative* data was: _____

_____

_____

b) I chose this method because _____

_____

c) An alternative method I could have used was: _____

_____

d) This might have been better because _____

_____

Study **Figure 1**, a map of pedestrian (people) flows (in any direction) in York, in northern England.

Key Pedestrians per 15 minutes

Less than 50

50–99

100–199

200–399

400–599

600+

Figure 1 Pedestrian flows in York

1.1 Using the data in the table below, complete the pedestrian flow map in **Figure 1**.

[2 marks]

Site	Number of pedestrians per 15 minutes
1	562
2	59

1.2 Describe the pattern of pedestrian flows shown in **Figure 1.**

[4 marks]

_____

_____

_____

_____

_____

_____

_____

1.3 Suggest **one** advantage of using proportional flow arrows, such as the ones in **Figure 1,** to show geographical information.

[2 marks]

_____

_____

_____

_____

1.4 Suggest **one** possible reason why the pedestrian flow data in **Figure 1** may not be accurate.

[2 marks]

_____

_____

_____

_____

1.5 Suggest **one** alternative technique that the student could have used to show pedestrian flow data in **Figure 1.**

[2 marks]

_____

_____

_____

_____

Study Figure 2, a photographic transect along roads in a part of central London.

Figure 2 Transect along roads in central London

1.6 Suggest **one** reason why the student has chosen the method of data presentation used in **Figure 2.**

[2 marks]

_____

_____

_____

_____

1.7 Suggest **two** limitations of using photographs, like those in **Figure 2,** as part of data presentation.

[4 marks]

Limitation 1 _____

_____

_____

_____

Limitation 2 _____

_____

_____

_____

As part of an enquiry collecting primary physical geography data, a student measured the percentage plant cover along a footpath. The results are shown in Figure 3.

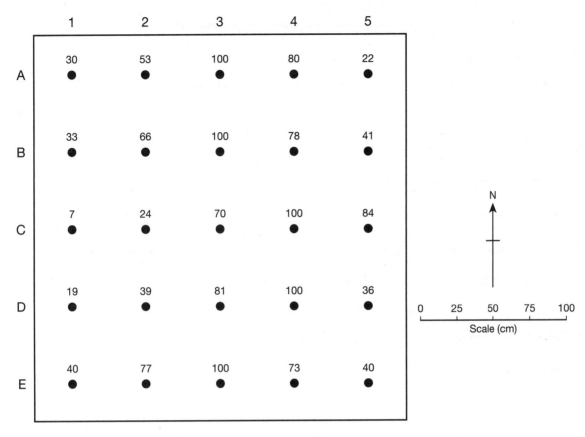

Figure 3 Percentage plant cover measured using point sampling

1.8   Complete the following table using the percentage cover data in **Figure 3**, column **4**.

[2 marks]

% plant cover column 4	Mean	86.2
	Median	
	Mode	

1.9   Calculate the range in **Figure 3**, row **C**.

[2 marks]

_____

_____

_____

_____

2.1   State the title of your fieldwork enquiry in which **physical** geography data were collected.

Title of fieldwork enquiry: _____

Explain **one** way in which you collated / summarised your primary fieldwork data.

[2 marks]

_____

_____

_____

_____

2.2   For **one** of your enquiries, explain one reason why you selected a specific data presentation method.

[2 marks]

Data presentation method: _____

Explanation: _____

_____

_____

_____

2.3   State the title of your fieldwork enquiry in which **human** geography data were collected.

Title of fieldwork enquiry: _____

a)   Justify the use of **one** type of graph used in your **human** geography enquiry.

[2 marks]

_____

_____

_____

_____

b)   Justify **one** cartographic technique used in your **human** geography enquiry.

[2 marks]

_____

_____

_____

_____

c) Assess the effectiveness of your data presentation methods that you used in this enquiry.

[6 marks]

_____

_____

_____

_____

_____

_____

_____

_____

_____

_____

_____

_____

2.4 With reference to data presentation methods used in **one** of your enquiries, explain to what extent these helped you to interpret your fieldwork data.

Title of fieldwork enquiry: _____

[9 marks]
[+ 3 SPaG marks]

_____

_____

_____

_____

_____

_____

_____

_____

_____

_____

_____

Continue your answer on another piece of paper.

Geographical Enquiry Strand 4

1 Describing, analysing and explaining your results.
2 Making links between different sets of data.
3 Using statistical techniques to understand the data.
4 Spotting anomalies in fieldwork data.

1 Describing, analysing and explaining your results

Describe and explain are commonly used terms. But analyis may be less familiar. It means making sense of what you have found out in your fieldwork data. It can involve using quantitative and qualitative techniques to help you to reach conclusions (see page 27).

Analysis has three stages – describing, explaining, and making links.

Describing and explaining

Describing means simply saying what the fieldwork data show, and is probably the first thing that you'd do with your data. Most likely, the data will be a table of results, a graph (like the one in Figure 1) or even a map.

Figure 1 A bar graph showing average scores from a shopping survey, where shoppers were asked to rate eight qualities of their local shopping centre (A to H) on a scale from 0 (Poor) to 5 (Excellent).

Explaining means that you look for reasons for what you have described. For example, you might have spotted that your data showed a pattern when you were describing them – such as pebble size getting smaller along a beach. So, you could try to explain this, perhaps using a process such as longshore drift.

If unsure, you might suggest reasons. For example, in Figure 1, you could suggest reasons why shoppers in the survey rated qualities C and D more highly.

Big idea

Data are often complex and need to be interpreted, so we can make sense of them. Sometimes data give surprising results – but just because we didn't expect them, we shouldn't assume they are wrong!

Exam hint!

Remember that the exam will never ask you to describe your own (familiar) fieldwork data, but could ask you to describe data collected by others (unfamiliar fieldwork data).

Exam hint

Learn some snippets of your data to quote when you're explaining what your fieldwork showed. A couple will do – you don't need to learn the whole lot.

Using the data in Figure 1, identify and describe:

a) the highest value or category _____

b) the lowest value or category _____

c) the difference between the highest and lowest values (called the **range**) _____

d) the mean (or average), and which categories are average, above or below average _____

_____

2 Making links between different sets of data

Being able to link one set of results to another can be difficult. For example, you might use a questionnaire to find out how people travelled to go shopping in a town centre, and a land use map. You might find that people questioned in one part of a shopping centre travelled mostly by train – checking its location on a map might show it was near the train station!

Using a spider diagram as a way of showing links can work well, as Figure 2 shows. Why not try this with your fieldwork results?

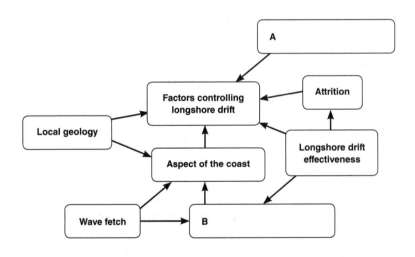

Figure 2 A spider diagram which shows possible factors affecting longshore drift for a stretch of coastline

Study Figure 2.

a) Identify what might go (i.e. what factors) in boxes **A** and **B** on the diagram, and write these in.

b) Add more 'spokes' and text boxes to the diagram to explain the possible factors affecting longshore drift.

3 Using statistical techniques to understand the data

Most analysis of fieldwork data involves some statistical techniques. You may be asked to explain how they are used, as well as doing some calculations linked to unfamiliar data that is given to you in the exam. Complete the activity below to check you know some commonly used techniques.

Over to you

Match the techniques in the left hand column with the definitions in the right hand column. One has been done to start you off.

Technique	Definition
Median	Working out an increase or decrease as a percentage.
Range	The middle value in a ranked set of data (4th of 7, 11th of 21 etc.).
Mode	Dividing a list of numbers into four equal groups (or quarters) – two above and two below the median.
Quartiles	The number that appears most commonly in a data set (this would be the highest bar on a bar chart).
Percentage change	A line indicating the general course or tendency of something, e.g. a set of points on a graph.
Trend line (also known as a 'line of best fit')	The difference between the highest and lowest values – subtract one from the other.

Figure 3 Statistical techniques that you need to know

4 Spotting anomalies in fieldwork data

Anomalies are unusual data that don't fit the rest of the results. You might spot them on a scattergraph or, for example, river flow data might increase unexpectedly at one point. Data could be different because someone collected them wrongly, or data could be correct but just different from the rest. You are not expected to know what caused the anomalies, but just to be able to spot them. Work through *Over to you* on page 37 to spot the anomalies in Figure 4.

Site No	Average velocity (m/s)	Average sediment size (mm, measured along the longest axis)
Site 1	0.22	23
Site 2	0.26	42
Site 3	0.23	108
Site 4	0.43	32
Site 5	0.33	45
Site 6	0.87	15
Site 7	0.45	25
Site 8	0.42	22

Figure 4 Velocity and sediment data from eight recording points along a river course

Over to you

1   Draw and label two graph axes in the space below, one for each set of data in Figure 4.

2   Plot each set of fieldwork data from eight points along a river shown in Figure 4.

3   Identify the two anomalies on the graph.

   Anomaly 1 _____

   Anomaly 2 _____

Get exam ready

In the exam, be ready to explain how you analysed data in your enquiry. Do this by completing the details below.

1a)   One method I used to analyse my *quantitative* or *qualitative* data was: _____

   _____

b)   I chose this method because _____

   _____

c)   An alternative method I could have used was: _____

   _____

d)   Which might have been better because _____

   _____

As part of a river study, a student measured pebble sizes (along the longest axis) at three places on a river - sites, A, B and C. Their results are shown in Figure 1.

Site A upstream	Site B middle section	Site C downstream
95	24	10
68	19	12
49	45	32
90	29	34
82	18	21
86	48	67
56	55	12
80	45	19
49	35	19
69	13	12
68	15	18
57	1	22
70	19	15
59	21	9

Figure 1 Pebble measurements in mm

1.1 Using the data in **Figure 1**:

a) Calculate the median sediment size for **Site A**.

[1 mark]

Median = _____

Space for calculation (you may use a calculator if you wish): _____

_____

_____

b) Circle the anomaly in the column for **Site B**.

[1 mark]

c) Calculate the interquartile range of the pebble size data for **Site C**.

[2 marks]

Interquartile range = _____

Space for working: _____

_____

_____

1.2   Using **Figure 1**, compare the differences in pebble size between **Sites A** and **C**.

[4 marks]

_____

_____

_____

_____

_____

_____

_____

1.3   Using **Figure 1**, suggest reasons for the differences in pebble size between the three sites.

[4 marks]

_____

_____

_____

_____

_____

_____

_____

As part of an enquiry collecting primary human geography data, a group of student recorded the number of shops and services in towns and villages of different population sizes. Their results are shown in a graph, Figure 2 (see page 40).

1.4   Using the data in **Figure 2**, calculate the **range** in the number of shops and services. Show your calculations.

[2 marks]

Range = _____

Space for calculation: _____

_____

_____

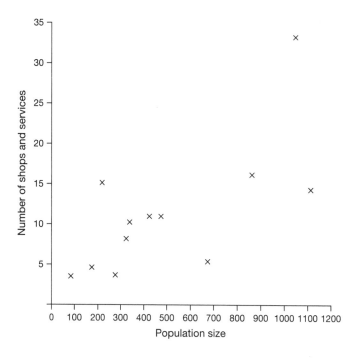

Figure 2 Number of shops and services for towns and villages of different sizes

1.5 Use the data in **Figure 2** to:

a) Identify the relationship between population size and number of shops and services.
Place a cross next to the word that describes the relationship in column **A** in the table.

[1 mark]

b) Place a cross next to the word that describes the strength of the relationship in column **B**.

[1 mark]

Relationship	Column A	Strength of relationship	Column B
Positive		Strong	
Negative		Weak	
Random		No relationship	

1.6 Suggest **one** reason for the pattern shown between population size and the number of shops and services in **Figure 2**.

[2 marks]

_____

_____

_____

_____

As part of a fieldwork trip to collect human geography data, a student photographed two different areas of Barcelona, a large city in Spain. Two examples of their photographs are shown in Figure 3.

Figure 3 Barcelona, Spain

1.7 Suggest how annotations could be used to analyse the information in each photograph.

[2 marks]

_____

_____

_____

_____

1.8 Explain **two** limitations of the analysis of qualitative data such as photographs.

[4 marks]

Limitation 1 _____

_____

_____

_____

Limitation 2 _____

_____

_____

_____

2.1 State the title of your fieldwork enquiry in which **physical** geography data were collected.

Title of fieldwork enquiry:

_____

Explain **one** method that you used to analyse your **primary** fieldwork data.

[2 marks]

_____

_____

_____

_____

2.2 Justify the use of **one** statistical technique that you used to analyse data in your **physical** geography enquiry.

[3 marks]

_____

_____

_____

_____

_____

_____

2.3 State the title of your fieldwork enquiry in which **human** geography data were collected.

Title of fieldwork enquiry:

_____

Explain the geographical links between any two sets of data that you collected in your **human** geography enquiry.

[3 marks]

_____

_____

_____

_____

_____

_____

2.4 Assess the effectiveness of one technique you used to analyse your fieldwork
data in your **human** geography enquiry.

[6 marks]

_____

_____

_____

_____

_____

_____

_____

_____

_____

_____

_____

2.5 With reference to **one** of your fieldwork enquiries, suggest how you could have improved
the analysis of your data.

Title of fieldwork enquiry:

_____

[9 marks]
[+ 3 SPaG marks]

_____

_____

_____

_____

_____

_____

_____

_____

_____

Continue your answer on another piece of paper.

Geographical Enquiry Strand 5

1 Drawing an evidenced conclusion for your enquiry question (or hypothesis).

How do I write a conclusion?

To write a conclusion, you need to go back to your enquiry question. Think about the following:

- What have you found out? Can you answer the original question for your fieldwork enquiry?
- Which of your primary data most strongly support your conclusion?
- Have any of your secondary data confirmed that your conclusions are right?
- Would any other evidence help to make your conclusions stronger?

Figure 1 is designed to help you think about your conclusion and work out what it is that you have found out in your fieldwork enquiry.

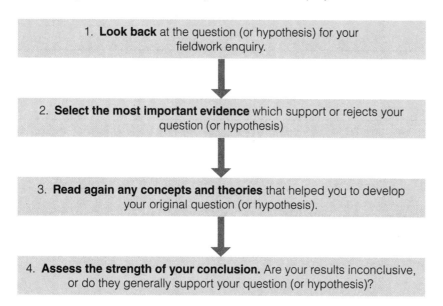

1. **Look back** at the question (or hypothesis) for your fieldwork enquiry.

2. **Select the most important evidence** which support or rejects your question (or hypothesis)

3. **Read again any concepts and theories** that helped you to develop your original question (or hypothesis).

4. **Assess the strength of your conclusion.** Are your results inconclusive, or do they generally support your question (or hypothesis)?

Figure 1 A sequence of processes to help you write your conclusion

Big idea

Conclusions are summaries of what you have found out. To write a good conclusion, look back at your original question or hypothesis, and think how far your data allow you to answer it.

Exam hint

Try and learn your conclusions to each piece of fieldwork that you did – no more than a single sentence for each!

Over to you

Complete the following:

a) The enquiry question for my physical / human enquiry was _____

_____

b) The answer I can give to my question is _____

_____

c) The most important evidence which supports my conclusion is _____

_____

d) I think that my conclusion is *strong / weak (delete as necessary)* because _____

_____

1 Drawing an evidenced conclusion for your enquiry question (or hypothesis)

Evidenced conclusions are those where you use data from your fieldwork to state a geographical outcome. You should therefore refer to the results of your fieldwork wherever you can – and be prepared to quote a couple of examples of your data in the exam to illustrate what you are trying to say.

However! You might have some results that don't fit the pattern of the rest because they are **anomalies** (see Strand 4 page 36). You need to think about these, and any other results that don't fit.

- What might have caused these? For example, if it's been a dry summer, stream levels may be low and velocity data may be hard to collect.
- Do any anomalies make your conclusion less certain? If so, why?
- Can any anomalies be ignored? Again, consider why.

Over to you

Study Figure 2 which shows pedestrian counts at ten sites from west to east across a shopping centre. Decide which of the following statements represent what is shown in Figure 2. If they can be improved using evidence from Figure 2, re-write them in the spaces below. And remember – always quote examples of data!

a) There is a decline in the number of pedestrians in the shopping centre from west to east.

Rewrite: _____

_____

b) There are more pedestrians in the west than the east except for in a few places on the way.

Rewrite: _____

_____

c) Pedestrian flow data generally declines, though there seems to be an anomaly around sites 5-7 where the number rises.

Rewrite: _____

_____

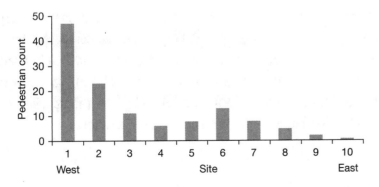

Figure 2 Pedestrian counts at ten sites from west to east across a shopping centre

What if my conclusions aren't as I expected?

Sometimes things don't always work out as you imagine. It doesn't matter if your conclusions are different to what you might have expected. Sometimes, this is referred to as 'messy' fieldwork! In fact, it might be an advantage, as it gives you more to write about in explaining links between your data, and the idea or theory you were trying to test.

Looking beyond

The final part of a conclusion is about the wider geographical significance of your study.

- Why might your study be important?
- Would any of your results be useful to other people or organisations such as local businesses or councils?
- Have you found out something that would apply to other places – e.g. would all rivers be like yours, or all town centres show similar results?

Writing about your conclusions in the exam

Look at the extracts in Figures 3 and 4 from two students answers to the exam question below. Some guidance is given around the exam question to help you understand what is needed in a good answer.

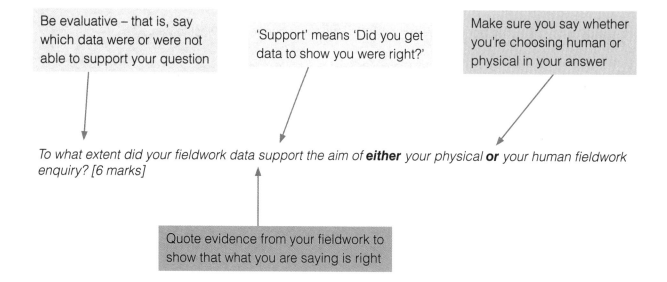

Be evaluative – that is, say which data were or were not able to support your question

'Support' means 'Did you get data to show you were right?'

Make sure you say whether you're choosing human or physical in your answer

*To what extent did your fieldwork data support the aim of **either** your physical **or** your human fieldwork enquiry? [6 marks]*

Quote evidence from your fieldwork to show that what you are saying is right

Student A

Our **human** enquiry was to see whether rebranding had been successful in Carlisle's shopping centre. We found overall it was successful based on our pedestrian flow data (because we compared data from ten years ago with present values), for the majority of locations we looked at in the town. Numbers of shoppers were up, as we predicted in our aims. These data were therefore useful in reaching our conclusion. But environmental quality was less certain as it varied a lot and did not match with our hypothesis that environmental quality would be higher near the areas of regeneration.

Figure 3 A student answer to the question above. Look for the three elements in the student answer which are highlighted to match the exam question guidance above. This student got 6 marks.

Student B

We found that the gradient of the beach is mainly due to the local geology and the fetch. Some winter storm waves are also having an effect. We found that our data presentation was successful, allowing us to see where the different sized stones were found at Compton Bay. There were some large stones in different places but most were at the western end of the beach where storms were worst. Some results were different from our expected theories, e.g. the fact that the angle of the beach varied a lot. We concluded that the theory of longshore drift didn't seem to work on this beach except at the western end.

Figure 4 An answer to the same question from Student B

 Over to you

1   Study the answer to the exam question from Student B.
Use highlighters to identify places where a) the student is evaluating, b) providing evidence, and c) where the student thinks they have support (or not) for their hypothesis.

2   Read the mark scheme below.
a) What level would you say Student B's answer is? _____
b) How should Student B improve their answer?_____

_____

> **Level 3** – The student makes detailed links between the aims of the enquiry, using evidence, and writes in an evaluative way which show clearly how strong the conclusions are.
> **Level 2** – The student makes some links between the aims of the enquiry, using some evidence, and shows in a general way how strong the conclusions are.
> **Level 1** – The student makes limited links between the aims of the enquiry but with little evidence, and does not show how strong the conclusions are.

✓ **Get exam ready**

In the exam, be ready to explain how you drew conclusions in your enquiry.
Do this by completing the details below.

1a) One conclusion I can draw from my **physical / human** fieldwork is: _____

_____

b) This conclusion is firm / tentative (*cross out one*) because: _____

_____

c) The data which support my conclusion most strongly are: _____

_____

d) The anomaly or data which makes my conclusion less certain is: _____

_____

e) My conclusion would be even firmer if I could find out: _____

_____

Geographical Enquiry Strand 5

Exam-style questions: Unfamiliar Fieldwork

Study Figure 1, an extract from an interview to investigate people's attitudes towards the increasing number of coffee shops on their local high street.

> I use local coffee shops a lot. Perhaps too much – I spend all my money on expensive coffees! It's silly really as I can make a coffee at home for a fraction of the price that the coffee shops charge. There are lots of people like me who pop in for a takeaway coffee. We are creating the need for more coffee shops on the high street. Have we reached 'peak coffee'? Probably not. With more shops closing, coffee shops could fill vacant properties.

Figure 1 Extract from interview

1.1 Explain **one** conclusion that can be drawn from the evidence provided in **Figure 1**.

[2 marks]

1.2 Suggest **two** further sources of information that could help to strengthen your conclusion based on **Figure 1**.

[4 marks]

Study Figure 2, photographs taken during an investigation into environmental quality in different areas of East London.

Figure 2 Investigating environmental quality in East London

1.3   Explain **one** conclusion that can be drawn about environmental quality from the evidence provided in **either** photograph **A** or **B**.

[2 marks]

Circle the photograph chosen: **Photo A   Photo B**

_____

_____

_____

_____

Study Figure 3, which is river data collected by a group of geography students.

Channel variable	Site 1	Site 2	Site 3	Site 4	Site 5
Width (m)	0.42	0.52	0.78	0.85	1.10
Depth (m)	0.10	0.13	1.6	0.18	0.21
Cross-sectional area (m²)	0.04	0.07	1.24	0.15	0.23
Velocity (m/sec)	0.45	0.47	0.56	0.55	0.51
Discharge (m³/sec)	0.02	0.03	0.70	0.08	0.12

Figure 3 River data

1.4   Explain **one** conclusion that can be drawn from the depth data in **Figure 3**.

[2 marks]

_____

_____

_____

1.5   Comment on the strength of conclusions that can be drawn from the data in **Figure 3**
about the following hypothesis:

Rivers change at a uniform rate as they flow downstream.

[4 marks]

_____

_____

_____

_____

_____

_____

_____

_____

**Study the diagram in Figure 4. It shows the characteristics of different types of people
who use beaches, from surveys carried out by students.**

A nice cup of tea	Unspoilt wilderness	Arcades and Rollercoasters	Buckets and spades
CHARACTERISTICS: Walks and places to eat. Older couples or singles. WANT: good places to eat, clean beaches, places of interest.	CHARACTERISTICS: Enjoying nature and wildlife. Wealthier couples and families. WANT: parking, limited crowds, scenery, unspoilt places.	CHARACTERISTICS: Funfairs and arcades. Families with older children. WANT: low cost entertainment, lots of affordables places to eat, clean beaches.	CHARACTERISTICS: Playing on the beaches. Families with younger children. WANT: clean beaches, sunshine and wet-weather alternatives.

Figure 4 Beach users

1.6 Explain **one** conclusion that could be drawn from **Figure 4**.

[2 marks]

_____

_____

_____

_____

1.7 Explain **one** limitation of a diagram like that in **Figure 4** in trying to draw conclusions.

[2 marks]

_____

_____

_____

_____

A group of geography students decided to collect information about the gradient (slope) of a local beach as part of a survey about beaches. After taking two measurements at either end of the beach (east and west) they developed the following conclusion:

The beach profile is steeper at the eastern section compared to the west section.

1.8 Explain why the reliability of their conclusion may be poor.

[2 marks]

_____

_____

_____

_____

1.9 Suggest an alternative conclusion that they could reach.

[2 marks]

_____

_____

_____

_____

Geographical Enquiry Strand 5

Exam-style questions: Familiar Fieldwork

2.1 State the title of your fieldwork enquiry in which **physical** geography data were collected.

Title of fieldwork enquiry:

_____

Explain to what extent your **secondary data** helped to support your conclusions.

[6 marks]

_____

_____

_____

_____

_____

_____

_____

_____

_____

_____

2.2 Explain to what extent a **geographical theory** or **concept** helped you to draw your conclusions in your **physical** geography enquiry.

[6 marks]

_____

_____

_____

_____

_____

_____

_____

_____

_____

_____

2.3　　State the title of your fieldwork enquiry in which **human** geography data were collected.

Title of fieldwork enquiry:

_____

Asses the strength of the conclusions that you were able to draw from your fieldwork.

[6 marks]

_____

_____

_____

_____

_____

_____

_____

_____

Continue your answer on another piece of paper.

2.4　　With reference to **one** of your fieldwork enquiries, assess the extent to which your conclusions matched your expectations at the start of your enquiry.

Title of fieldwork enquiry:

_____

[9 marks]
[+ 3 SPaG marks]

_____

_____

_____

_____

_____

_____

_____

Continue your answer on another piece of paper.

Geographical Enquiry Strand 6

1   Identifying problems with data collection.
2   Spotting limitations in the data.
3   Thinking of ideas for other data that might be useful.
4   Judging how far your conclusions were reliable.

1 Identifying problems with data collection

Errors in data collection can raise problems. What matters is whether the errors affect your results, and therefore the reliability of your conclusions.

Look at the data in Figures 1a and 1b. In each set of data, there is at least one likely error. Circle where you think these might be. The likely source of error is one of three types:

- **Measurement error** – mistakes made when collecting data, e.g. misreading a thermometer.
- **Operator error** – differences in results collected by different people, e.g. when individuals give different scores for the same thing.
- **Sampling error** – local differences, e.g. where one area gives very different results to another.

However, be careful not to judge too quickly – some scores are **intended** to bring out differences, such as scores which rate people's opinions.

Big idea

Evaluating your enquiry is the final part of the enquiry process. It's a chance to think about things like sample size, timings, and problems with equipment, so that you can judge whether these affected your results and conclusions.

	Stone length (mm)	Stone width (mm)	Beach gradient (degrees)
Site 1	104	23	2
Site 2	210	34	6
Site 3	120	67	4
Site 4	133	92	27
Site 5	12	45	4

Figure 1a Data from a beach where stone or pebble length / width and beach gradient were measured

	Housing Quality Score 1 (poor) to 20 (high)		
	Scorer 1	Scorer 2	Scorer 3
Area A	5	7	6
Area B	2	2	1
Area C	15	17	7
Area D	17	17	19

Figure 1b Data from a town where housing quality was scored

Over to you

1   Draw a line to match the problems with the likely cause of error in the table below.
2   Rank the problems in terms of their effect on the reliability of results, where 1 = most serious.
3   List any problems that you experienced with your data recording. _____

_____

Problem	Rank	Likely cause of error
Limited sample size		A student misread the gun clinometer
Wrong time (season)		The questionnaire did not have sufficient categories for different ages
Equipment problems		The river was too low to measure
Recording sheet errors		The beach sediments were only measured at three places
Wrong time (during the day)		It was raining when the students tried to find people to interview in the town
Operator error		The GPS unit ran out of batteries

2   Spotting limitations in the data

Limitations in your data collection can be linked to a number of reasons. It's important to identify and explain these, because any weaknesses in your fieldwork method or design could lead to possible inaccuracies, and therefore create unreliable conclusions.

 Exam hint!

Don't just expect exam questions to ask about problems or limitations of your data. You will most likely be asked to explain how any problems affected the **reliability** of your enquiry.

 Over to you

1   Look at the extract in Figure 2 from a students' answer to the following exam question.

Discuss the ways in which limitations of your data collection affected the reliability of your results. (6 marks)

> We found that the high tide created a problem in getting access to the lower part of the beach to measure gradient. This limited our data collection because we were unable to measure the full gradient between the high and low water points at the beach on the day.

Figure 2 Extract from student answer

2   Highlight in different colours the places in which the student in Figure 2 discusses a) problems and b) any limitations in their data collection.

3   Explain what makes this a high quality answer.

3 Thinking of ideas for other data that might be useful

Additional data could help to improve the reliability of your conclusions. There are two things you could do - collect more of the same type of data, or find different data, as Figure 3 shows.

More of the same type of data	Additional and new data
• Doing more measurements at different times of the day. • Increasing the sample size with a second visit.	• Secondary data from a similar study done at a different time. • Looking up new maps to help understand change.

Figure 3 Gaining more data can help your enquiry

Over to you

1   Complete the following to explain how more data might have helped your **physical** or **human** fieldwork enquiries.

Fieldwork chosen (delete as appropriate) Physical / Human

a) **Having more of the same type of data**

(i)   Doing more measurements at different times of the day.

_____

_____

_____

(ii) Increasing the sample size with a second visit.

_____

_____

_____

b) **Additional and new data**

(i)   Secondary data from a similar study done at a different time.

_____

_____

_____

(ii) Looking up new maps to help understand change.

_____

_____

_____

4   Judging how far your conclusions were reliable

Reliability is a tricky idea, though you might have come across it before,
e.g. in science. But there are other terms that will help you understand this
last stage of the enquiry process, and which will be useful in writing-up your
evaluation as shown in *Over to you* below.

Over to you

Figure 4 shows six terms, or concepts, which you can use to work out the reliability
of the data that you collected. Draw a line to link each term with the correct definition.

Term/concept	Definition
Accuracy	The difference between the result you found and its true value if it were done using professional equipment.
Reliability	Data which are unusual and differ from the rest.
Error	How suitable the methods of collecting data are for the question that you are investigating.
True value	How close the data you collected are to their real value.
Anomalies	The data that would be obtained in an ideal measurement using methods that are as accurate as possible.
Validity	How consistently the same data that you obtained would be repeated if you went back at another time.

Figure 4 Reliability – terms and definitions

Get exam ready

In the exam, be ready to evaluate any methods you used, data you collected, or conclusions
that you drew in your enquiry. Do this by completing the details below for **one** of your enquiries.

1a) A possible source of error in my **physical / human** investigation was: _____

_____

_____

b) The ways in which these errors might have affected my results are: _____

_____

_____

c) My conclusions might be affected because: _____

_____

_____

Study **Figure 1** which shows three pieces of fieldwork equipment which may be used in geography fieldwork enquiries.

Figure 1 Fieldwork equipment: **A** Video camera; **B** Clinometer; **C** Plastic metre ruler

1.1 Select **one** item of equipment in **Figure 1**. Suggest why the results students obtain from it might be inaccurate.

[2 marks]

Item selected: _____

Why results might be inaccurate: _____

_____

_____

_____

1.2 For **one** item of equipment in **Figure 1**, suggest one way of improving students' accuracy.

[2 marks]

Item selected: _____

One way to improve accuracy: _____

_____

_____

_____

Study **Figure 2**, a proposed design for sampling data in an investigation into how microclimate changes from the centre of a woodland to its outskirts.

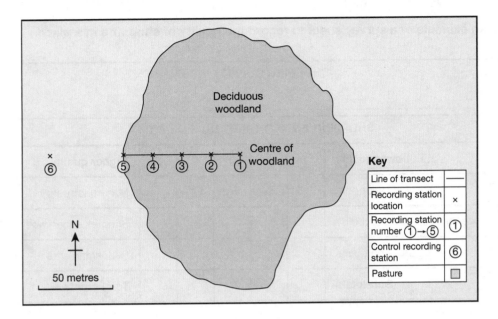

Figure 2 Design for sampling data

1.3 Suggest **one** problem with the design of the fieldwork sampling strategy shown in **Figure 2**.

[2 marks]

_____

_____

_____

_____

1.4 Explain **one** way of improving the reliability of the design in **Figure 2**.

[2 marks]

_____

_____

_____

_____

1.5 Using **Figure 2**, assess how far any conclusions from this enquiry might be accurate.

[2 marks]

_____

_____

_____

_____

Geographical Enquiry Strand 6

Exam-style questions: Unfamiliar Fieldwork

Study Figure 3, an example of a survey sheet to record the quality of shopping in a place.

SHOPPING QUALITY INDEX						
Street name:						
SHOPPING ENVIRONMENT QUALITY INDEX						
Lower quality	1	2	3	4	5	Higher quality
Few shoppers						Many shoppers
Shop exterior – run down and dirty						Shop exterior – well maintained
Traffic, parking, no crossing points						Pedestrian area
Sub-totals						Total points =

Figure 3 A shopping quality index survey sheet

1.6   Using **Figure 3**, identify **two** limitations to this shopping quality index survey sheet.

[4 marks]

Limitation 1 _____

_____

_____

_____

Limitation 2 _____

_____

_____

_____

1.7   Suggest **one** way in which the index in **Figure 3** could be altered to improve its reliability.

[2 marks]

_____

_____

_____

_____

1.8   Explain how much the time of day or season might affect the results of a survey like the one in **Figure 3**.

[4 marks]

_____

_____

_____

_____

_____

_____

_____

_____

1.9   Suggest **one** alternative source of data that could be used instead of **Figure 3**.

[2 marks]

_____

_____

_____

_____

1.10   Suggest **one** secondary source of information that could be used to investigate the quality of an urban or rural area.

[2 marks]

_____

_____

_____

_____

Geographical Enquiry Strand 6

Exam-style questions: Familiar Fieldwork

2.1 State the title of your fieldwork enquiry in which **physical** geography data were collected.

Title of fieldwork enquiry:

_____

Explain **one** problem that you encountered in your data collection.

[2 marks]

_____

_____

_____

_____

2.2 Explain how the time of day or year of your **physical** fieldwork enquiry may have affected your results.

[3 marks]

_____

_____

_____

_____

_____

_____

2.3 For **one** of your enquiries, explain how your sample size may have affected the reliability of your conclusions.

Title of fieldwork enquiry:

_____

[2 marks]

_____

_____

_____

_____

2.4 Referring to **one** fieldwork investigation you have carried out, assess the range of additional data that could be used to improve your results.

Title of fieldwork enquiry:

_____

[9 marks]
[+ 3 SPaG marks]

_____

_____

_____

_____

_____

_____

_____

_____

_____

Continue your answer on another piece of paper.

2.5 Referring to **one** fieldwork investigation you have carried out, assess the problems that you faced with your methods of data collection.

Title of fieldwork enquiry:

_____

[9 marks]
[+ 3 SPaG marks]

_____

_____

_____

_____

_____

_____

_____

_____

_____

Continue your answer on another piece of paper.

UNIVERSITY PRESS

Great Clarendon Street, Oxford, OX2 6DP, United Kingdom

Oxford University Press is a department of the University of Oxford.
It furthers the University's objective of excellence in research, scholarship,
and education by publishing worldwide. Oxford is a registered trade mark
of Oxford University Press in the UK and in certain other countries

British Library Cataloguing in Publication Data

Data available

ISBN 978-019-842662-2

Kindle edition ISBN 978-019-842663-9

10 9 8 7 6 5 4 3 2 1

Printed in Italy by L.E.G.O SpA

Acknowledgements

The publisher and authors would like to thank the following for
permission to use photographs and other copyright material:

Cover: watchara/Shutterstock; **p6, 8, 11, 18, 30, 41, 49, 58:** David
Holmes; **p9:** Bob Digby; **p18(bl):** With kind permission of Environmental
Systems Research Institute, Inc.

Artwork by Aptara Inc.

Every effort has been made to contact copyright holders of material
reproduced in this book. Any omissions will be rectified in subsequent
printings if notice is given to the publisher